Practical Data Cleaning

19 Essential Tips to Scrub Your Dirty Data

Copyright

Practical Data Cleaning (5th Edition)

By Lee Baker

Copyright 2019 Lee Baker

Paperback Edition

Thank you for purchasing **Practical Data Cleaning.**

You are welcome to share it with your friends.

This book may be reproduced, copied and distributed for non-commercial purposes, provided the book remains in its complete original form.

If you enjoyed this book, please feel free to **contact the author**[†] for content feedback, permission questions or to discover other works by this author.

Thank you for your support.

***

[†] https://www.chi2innovations.com/contact/

Contents

Preface

Introduction

Chapter 1: Data Collection

Chapter 2: Data Cleaning

Chapter 3: Data Codification & Classification

Chapter 4: Data Integrity

Chapter 5: Work Smarter, Not Harder

About The Author

Leave a Review

Preface

The success of the first 4 editions of **Practical Data Cleaning** has been overwhelming and has come as a bit of a shock – I had no idea it would be so popular!

Firstly, a huge **thank you** to all of you that took the time to read it. I hope you found it useful and got something of real value out of it.

Secondly, I wish to thank all of you that have interrupted your busy schedules to get in touch. Your words of support have touched me and inspired me to continue writing.

Of all of the feedback I've received, by far the most common request is for me to cover the subject material in much greater depth, so I've created **The Hive** – an online portal where you can take video courses about Data Science – drawing on some of my 20+ years of experience.

To celebrate the launch of **The Hive**, I've created a **Practical Data Cleaning video course** to accompany this book. The video course goes into more detail than the book, and you can get it completely **FREE**!

There are lots of courses available in **The Hive** and many more planned, and you can get **FREE ENTRY** right now!

I hope you'll take a few moments to check out our video courses, and I look forward to seeing you there!

You can get access to the **Practical Data Cleaning video course** right now by following this link:

http://bit.ly/PDCHive

Introduction

Don't Panic !!!

***

Data cleaning is a waste of time.

If the data had been collected properly in the first place there wouldn't be any cleaning to do, and you wouldn't now be faced with the prospect of weeks of cleaning to get your dataset analysis-ready.

Worse still, your boss won't understand why your analysis report isn't on his desk yet, a mere 48 hours after he's asked for it. Bless him, he doesn't understand – he thinks that cleaning data is just about clicking a few buttons in Excel and – ta da! – it's all done. Even a monkey can do that, right?

And – for good reason – you won't get any help from statistics books either. Data is messy and cleaning it can be difficult, time-consuming and costly. Not to mention it's the least sexy thing you can do with a dataset.

Yet you've still got to do it, because, well, someone has to...

But it doesn't have to be so difficult. If you're organised and follow a few simple rules your data cleaning processes can be simple, fast and effective.

Not to mention fun!

Well, not fun exactly, just not quite as coma-inducing.

This book (now in its 5th Edition!) explains the 19 most important tips about data cleaning with a focus on understanding your data, how to work with it, choose the right ways to analyse it, select the correct tools and how to interpret the results to get your data clean in double quick time.

Now it doesn't matter whether you're a scientist or an entrepreneur, in academia or in business, if you're collecting data to try to answer some questions then **you need to understand the fundamentals**.

You'll likely spend a lot of time observing, measuring, counting, classifying and quantifying what you see, and once you've collected your data you're going to have to analyse it.

But let's not get too far ahead of ourselves...

Before you can get any answers you're going to have to:

- Collect
- Record & Store
- Clean & Classify

The textbooks tend not to dwell on the practical issues too much because, well, to be honest, it can get quite messy, but these are vitally important steps and you really do need to know how to do them properly if you're going to **get the most out of your data**.

So let's rewind to the beginning and see what we can do to get you off to a good start...

Here are 3 rules to start off with:

1. Don't Panic !!!
2. Start thinking about the data *before* you start collecting it
3. Make a personal vow to understand the basics of data

***

Learn More

There is a **Resources page** that accompanies this book, where you will find reviews and links to other **books**, **blogs**, **video courses** and all sorts of other useful stuff.

This page is updated regularly so you'll always know where to find the best the web has to offer.

Visit the resources page here:

- http://bit.ly/2Rb06Ds

Chapter 1: Data Collection

***

It may seem strange to start a book about data cleaning with a chapter on data collection, but when you think about it, if you collect data well, then your dataset will already be mostly clean and you'll have little work to do to make it analysis-ready.

In fact, such is the importance of good data collection practices, out of the 19 data cleaning tips in this book 11 of them are about data collection.

Ready to go?

OK, let's get into the first tip…

Tip #1: Record Data on Paper First…

So you've got your hypothesis (theory, idea or hunch). Once you've decided what data you need to collect, the first thing you should do is **design a paper-based form to store all your data** (assuming that at least some of your data is going to be recorded by hand).

Keep it simple, print it out, then manually record your data with pen and paper. One form per case/patient/customer/test-tube, etc.. It's also a good idea to do a test run collecting some real data. This way you'll get a much better understanding of whether the data you collect fits with your expectations (and I bet they don't!).

Physical Assessment:

Inprocessing BMI: _____

Current Weight: _____   Current BMI: _____

Heart Rate _____ BP _____ RR _____ T _____ LOC: Yes   No

I've had quite a bit of feedback about this tip, with most people disagreeing, suggesting that it's a redundant step and just wastes time.

Well, there are two really good reasons why I maintain that when data are collected from the Real World that they should not be entered directly into Excel:

1. Mistakes entered directly into Excel can be impossible to find and correct
2. Designing a paper-based form gives you the chance to think more deeply about your study

I'm not suggesting that mistakes don't happen when entering data on paper, but in my experience they are less frequent than when the data are entered directly to Excel – and if you've used a paper-based system you have an extra level of checks to find those errors.

It's the 2nd point that's more important, though. Planning a paper-based form means planning your study in depth. It's far more than a simple list – you get to group similar items together, deciding which ones are really needed and which are superfluous. Better still, when you're collecting data you'll often find that it doesn't always fit into your nice, neat boxes. On paper you can scribble notes and make corrections and additions, restructuring as you go. It's much harder to do that in a spreadsheet on-the-fly.

Tip #2: ...Then Transfer it to an Electronic Medium

We may be living in an electronic world, but ultimately you need a system where you (or anyone else) can **follow the data trail from beginning to end** and – more crucially – **from end to beginning**.

From time to time you WILL make a mistake with the data, so it is vitally important that you design a method that will let you spot and **rectify the mistake by going back through all the steps** until you find the error.

So now you have your data recorded on paper you need to transfer it into an electronic system. More than likely this will be either Microsoft Excel or Access.

In general, Excel is more common and easier to use, and has the added advantage that you can manipulate the data and do some simple analyses right there without having to export your data.

Most data is stored in Excel (in 7 years as a medical statistician I was only once given data in Access – all the other times it was in Excel), so we'll go with that from here on in…

Tip #3: Enter Your Data on a Single Worksheet Whenever Possible

Trying to sort your data when it is spread across multiple worksheets can lead to all sorts of problems, so try to avoid it whenever you can – **keep all your data on a single worksheet**.

Excel 2003 limits the number of usable worksheet rows and columns, and these limits are large enough for most datasets.

If you need higher limits you can use Excel 2010 or 2013.

Excel 2003 limits:
- 65,536 rows
- 256 columns

Excel 2010 and 2013 limits:
- 1,048,576 rows
- 16,384 columns

Tip #4: Use a Unique ID Column

You'll likely have to sort your data many times and by different columns, so you're going to need a way of restoring the original order.

Use column A as a unique identifier to insert consecutive numbers starting from 1. It may be simple, but it's very effective.

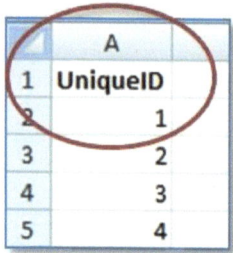

When you've put your Unique IDs into column A, go back to your original paper sheets and write the Unique ID there as well. Then sort your paper sheets by Unique ID.

Trust me – you'll thank me for this tip later…

Tip #5: One Column per Variable

Each variable should have… oh, hold on a minute, what's a variable?

Well, simply put, these are the things that **can change** or **can be changed** as part of your study. In short, these are all the pieces of information that you are observing, measuring, counting and collecting, like age, gender, distance, temperature, etc..

Where were we? Ah yes...

Each variable should have its own column, and each variable should correspond to just **one piece of information**.

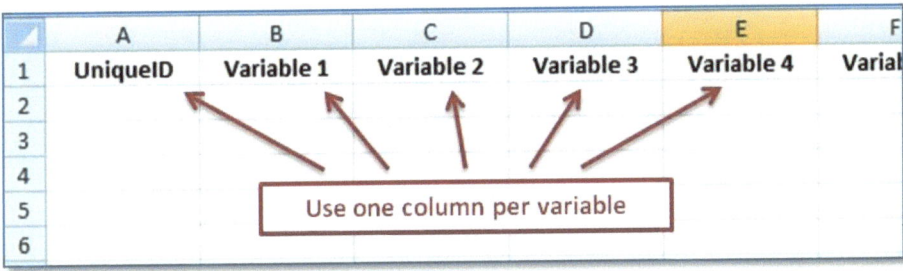

If you're entering the age of a patient, then just enter their age, don't enter their date of birth in the same column or cell.

If you want to record their age and DOB, then use 2 separate columns.

If you're recording a composite variable made up of 2 or more constituent parts, like Body Mass Index – made up of Height and Weight – then record them in separate columns.

You can always combine them into a single variable later – in a separate column of course...

Tip #6: Row 1 is the Variable Name

Eventually you'll need to analyse your data and you may need to export it to a statistical program.

The standard for pretty much all commercial stats programs is that **the first row is reserved for the name of the variable** and all other rows for the data.

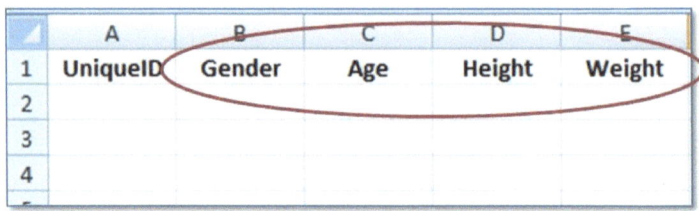

Don't be tempted to use rows 2, 3 and 4 as well as row 1 for the variable name – it might keep everything looking nice and tidy in Excel, but it will only create more work for you later.

If you have complicated names or additional information regarding the variable names, then you can use other worksheets to store this information.

Tip #7: Every Cell Should Have Something In It

What do empty cells tell you?

- waiting for more information?
- data not recorded?
- original data incorrect?

An empty cell is just a great big question mark and tells you nothing.

Worse still, incomplete datasets give reviewers a reason to whack you about the head with a metaphorical stick (and believe me they will – I've been there many times...).

So make sure that something is entered in every cell.

It is quite common to **use 'illegal' numbers as codes** to give you information, so where the entries for a variable can only be positive values (like age or height), we can use codes such as:

	A	B
1	My Variable Code	What It Really Means
2	-1	Data not recorded
3	-2	Waiting for lab
4	-3	Dave screwed it up, the idiot...
5		

Better still, **use letters a, b, c**, etc.. Using letter codes means that you can still do calculations on your numerical data, like the mean or median, without the answer being contaminated by codes.

It can be really useful to use codes like this, because you can then use Excel's Filter function to zone in on which cells need your attention, and you can get a count of how many cells containing each specific type of issue you've got in your dataset (then you can estimate how long your data cleaning will take, just in case your boss asks).

Tip #8: Keep Great Notes

When using codes you'll need to keep notes to tell you what the codes mean.

Keep the codes and notes in a different spreadsheet.

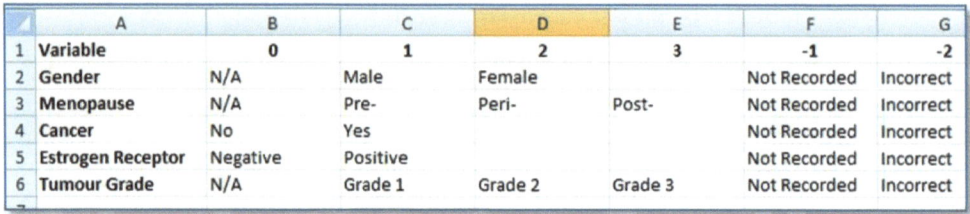

While we're on the subject, it's really important to:

KEEP GREAT NOTES !!!

You're likely not the only person that will ever work with this dataset, so **get used to writing stuff down**.

Explain what the project is all about, the questions you're trying to answer, why you're collecting this data and how you're going to get the answers you're looking for.

Explain how you measured things and under what conditions.

If more than one person is collecting data, then explain who, what, where, when, why and how.

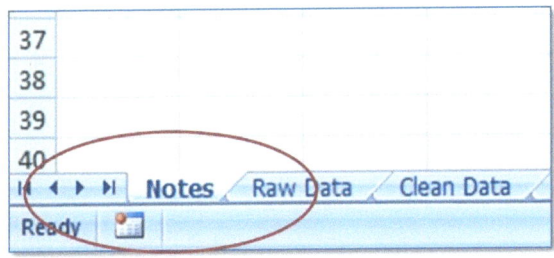

This will be the document that explains all the important stuff about your dataset, so **write it down**.

If there's too much information to comfortably put into an Excel spreadsheet, then a Microsoft Doc will be just fine – and keep it in the same folder as your dataset.

I guarantee that when it comes time to write your paper/thesis/report that you'll be glad you got into the habit of doing this – it'll all go so much quicker and easier.

Tip #9: Be Consistent

There's nothing worse than getting a dataset that takes a fortnight to clean because data entry has not been consistent.

By that I mean make sure that if the entry for a variable should be 'Positive', then make it 'Positive' and not some other variation (overleaf).

It's hard enough correcting speeling missteakes and typos without also having to correct things that were deliberately entered differently.

Restrict the number of people that can enter data to cut down on these issues, and make it clear what your **data entry standards** are.

I bet you didn't know it, but Excel has a Data Entry Form built in, so learn how to use it to reduce data entry errors. Oh yes, and learn how to use Data Validation too!

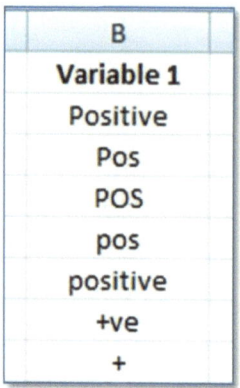

Tip #10: Don't Guess

Data should be entered as accurately as possible.

Don't guess, approximate, round up or down !!!

Enter the value exactly as registered on paper.

Don't do calculations in your head, on paper or in a calculator – you'll make mistakes which can be difficult, if not impossible, to spot later.

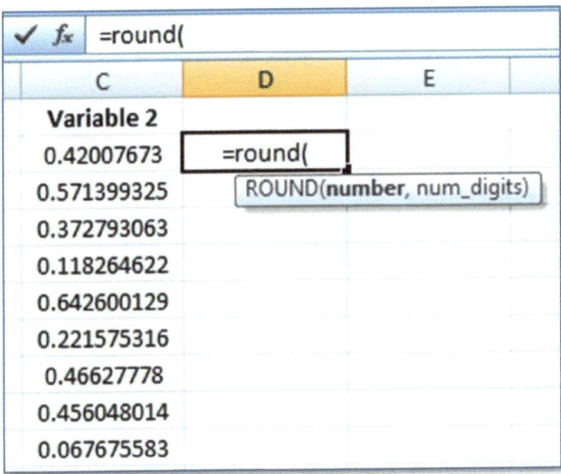

Then you can use Excel's built-in formulae to round your data to any number of decimal places or significant figures you like – consistently and safely.

Tip #11: Zero is a Real Number

Don't enter the number Zero into a cell unless what has been measured, counted or calculated results in the answer Zero.

I've often received datasets with lots of zeros and when I asked, the zeros meant 'I don't have data for this'.

The problem is that if you want to calculate something, like the mean, then all the zeros will be used in the calculation and you will get an **inaccurate answer** – or one that is **just plain wrong!**

> I see you're entering a zero.
>
> Are you sure this is really a zero or are you just storing problems for yourself later?

As with tip #7, if you don't have any data for a particular cell, don't leave it blank – enter a code (preferably a letter code) to tell you why you don't have the data.

***

Learn More

Just a little reminder about the **Resources page** that accompanies this book.

If you want to learn more about collecting, cleaning and processing data effectively and efficiently (and get even more **FREE stuff**), this is where you'll find it:

- http://bit.ly/2Rb06Ds

Chapter 2: Data Cleaning

***

If you've collected all your own data and you've been *very* careful you might just have a perfect dataset.

Well done!

Personally I've never seen a perfect dataset – it is the rarest of creatures.

Most likely you will have to clean your data before you can start to analyse it.

Yet again the textbooks will give you little practical advice here, so let's dive in and set a few ground-rules that will help you save time and keep your boss happy...

Tip #12: Make a Copy

You've got a 'raw' dataset that is essentially an electronic copy of all the paper-based data you have collected.

If you have made an entry error in the electronic copy you can always check back to the original paper copy.

When you move on to the data cleaning you're going to be changing the data and **you need to be able to undo any cleaning mistakes** you've made, and trust me – you're going to make a few.

So **create a duplicate worksheet** of your dataset.

Believe it or not, this is one of the most important steps in data cleaning.

Call the original one 'Raw Data' and the new one 'Cleaning In Progress' until you've finished cleaning, then you can change the name to 'Clean Data'.

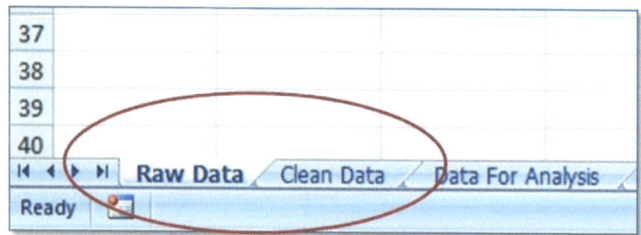

Once you have these data worksheets at various stages of preparation, you'll have a **chronological record** of your data cleaning processes. If you find an error in a later worksheet (and you will!), you can trace this error back through the worksheets until you find out how, when and why it was introduced. This is an excellent way of learning to build protocols into your processes to improve your methods for next time.

Oh yes – and make sure both worksheets have got the Unique ID column.

Tip #13: Clean Your Data in a Separate Worksheet

When cleaning an individual column of data you'll use a variety of different tools built into Excel, like 'Find And Replace'.

When you use 'Find And Replace' will it operate only on the selected column or on the whole worksheet?

Are you sure?

Really, really sure?

Do all of the in-built functions work in the same way?

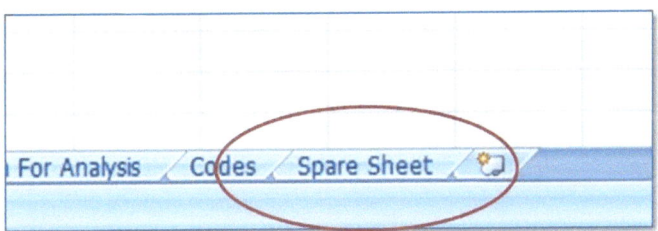

Get the answer wrong and you'll find that you've just introduced errors across your entire dataset with **no easy way to undo them** (hitting 'Undo' doesn't work here).

So when you want to clean a single column of data, **copy that column into a spare worksheet** and work on it there.

When you're done you can copy back, replacing the previous uncleaned column.

This is a process known as **ETL – Extract, Transform, Load**, and it may take you a little more time, but it's worth it – **mistakes can be very costly** and it pays to head them off at the pass.

Oh, I do hate that cliché…

Tip #14: Report Errors Back to the Original Source

It makes no sense cleaning your data if the same data has to be cleaned in exactly the same way time and time again. If you're using a shared dataset, such as a departmental database, make sure you **report back to the original source** any errors that you've found.

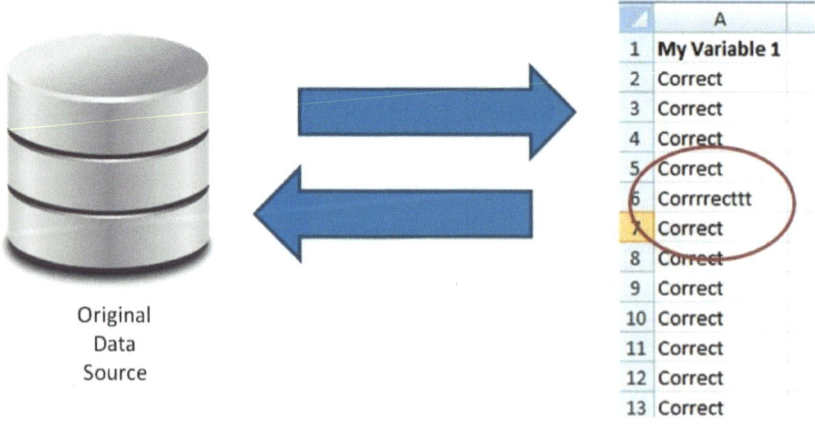

Then pester your data manager to update the database with the correct data. It's good for you, your boss, your colleagues and your karma. Better still, next time you have to analyse some more data from the same source you'll have a lot less cleaning to do.

Tip #15: Use Excel Functions to do the Hard Work...

Whenever possible, try not to clean data manually.

One of the biggest sources of spelling errors, typos and incorrect entries comes from manual entry, so why use the same method that got you into trouble in the first place?

Excel has a shed-load of **functions that can help with data cleaning**, so use them.

If you have a text-based column, use Excel's 'Remove Duplicates' function.

The result will be a list of all the items that you have in that column.

You can then use 'Find and Replace' to correct misspelled entries, including correcting entries with the wrong case, like 'case', 'Case' or 'CASE'.

Tip #16: ...And Use Excel Formulae to do the Even Harder Work

I cannot tell you how many weeks of my life I have lost – that I will never get back – trying to find the source of error that turn out to be a space at the beginning or end of the data in a cell.

You can't see it, but it's still there and it can wreak havoc when you start to do analyses.

Excel ignores spaces, so they can be incredibly difficult to detect, but other analysis and stats packages don't ignore them and they treat the entry as something different.

Spaces are the bane of my life!!!

So what to do?

Excel has a few different formulae that can be used to detect and trim spaces and other unwanted characters, like:

- TRIM()
- CLEAN()
- SUBSTITUTE()

So **learn how to do simple coding in Excel** and use these – and other – formulae.

I promise – it will definitely be time well spent!

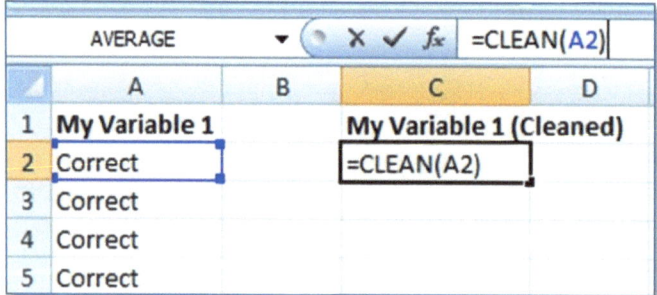

Make a personal commitment to learn how to use one new Excel formula every day. Within a month your colleagues will think you're an expert, and your boss will consider you an Excel savant. Maybe you'll get that next promotion after all…

***

Learn More

Still not visited the **Resources page** yet?

Well how about a little tempter? Visit the page to claim **FREE entry** to the Practical Data Cleaning video course that accompanies this book!

And if that's not enough, there's even more FREE stuff there to claim. Hop right over. You'll find it here:

- http://bit.ly/2Rb06Ds

Chapter 3: Data Codification & Classification

***

So you now have a perfectly clean dataset, but you still have some work to do before you start analysing it.

It's important that you **note what your codes mean** – after all, they're not a secret are they?

Say you've entered the data for a variable as 1, 2 or 3.

What does that mean?

- Small, Medium or Large?
- Pig, Sheep or Goat?

It matters because you shouldn't be expected to remember all the details of how, what and why you coded your data that way.

Tip #17: Keep a Code Sheet

Keep your codes in a separate worksheet and name it 'Codes'. For each column **make a note of what codes you've used** and what they really mean.

If you've used additional codes using 'illegal' entries such as negative numbers or letters, make a note of what they mean too.

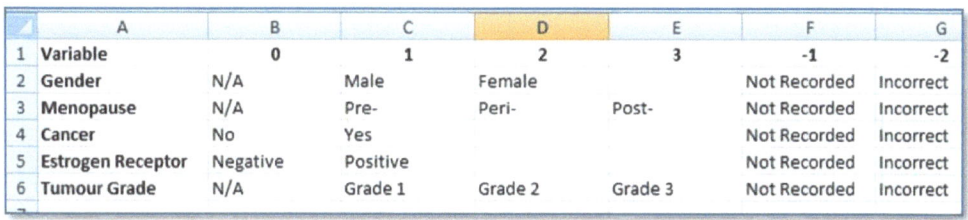

When you come back to the dataset after a couple of weeks away from it, you'll be glad you got organised like this.

You'll also make your boss, colleagues and local friendly statistician happy too, and **that's never a bad thing**...

Tip #18: Identify Your Data Types

When you get to the analysis stage you'll need to know your data types – Ratio, Interval, Ordinal and Nominal – so take a little time to decide which of these are appropriate for each variable, and note this down in your code sheet.

Not sure what these are? OK, then let's take a little step back.

There are two types of data:

- **Quantitative**
- **Qualitative**

Data is quantitative when it is measured with a ruler, jug, weighing scales, stop-watch, thermometer and so on.

It is qualitative when it is observed and placed into categories, such as gender (male, female), health status (healthy, sick), opinion (agree, neutral, disagree).

Quantitative and qualitative data can be sub-divided into four further classes of data:

- Quantitative (measured)
    - Ratio
    - Interval
- Qualitative (categorised)
    - Ordinal
    - Nominal

The difference between them can be established by asking just three questions:

1. **Ordered** – can some sort of progress be detected between adjacent data points or categories or can the data be ordered meaningfully?
2. **Equidistant** – is the distance between adjacent data points or categories consistent?
3. **Meaningful Zero** – does the scale of measurement include a unique, non-arbitrary zero value?

Nominal Data has the following properties:

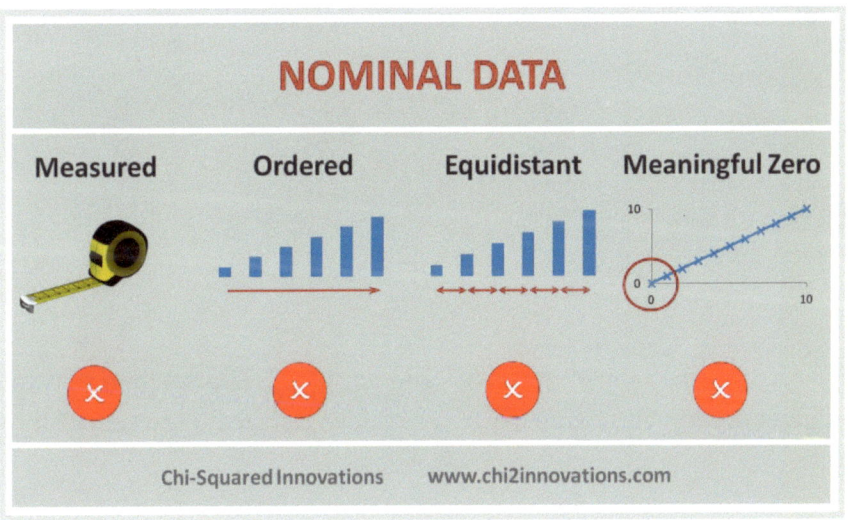

You can differentiate between nominal categories based only on their names, hence the title 'nominal' (from the Latin *nomen*, meaning 'name').

If the categories are descriptive (Nominal), like 'Pig', 'Sheep' or 'Goat', it can be useful to separate each category into its own column, such as Pig [Yes; No], Sheep [Yes; No], and Goat [Yes; No].

Ordinal Data has the following properties:

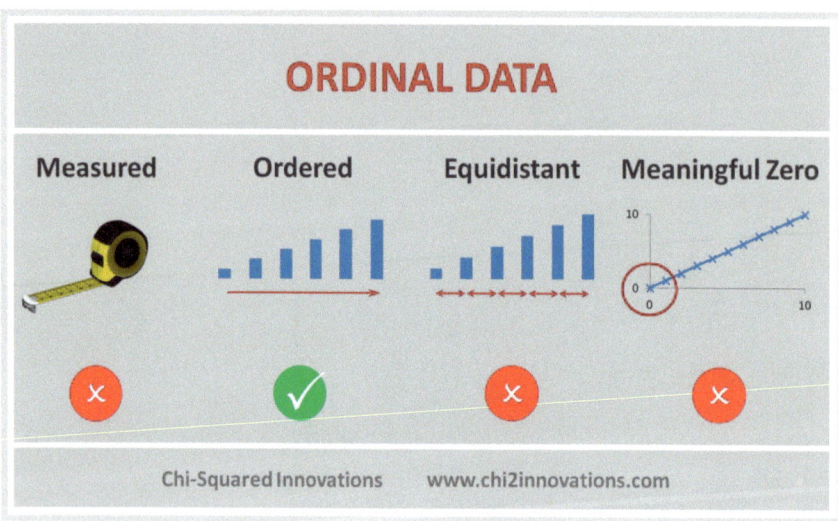

With ordinal data, the categories can be ordered; 1st, 2nd, 3rd, etc. (hence the name 'ordinal'), but there is no consistency in the relative distances between adjacent categories.

Mathematically, you can make simple comparisons between the categories, such as more (or less) healthy, agree more or less, etc., and since there is an order to the data, we can rank them and compute the median to find the central value.

Interval Data has the following properties:

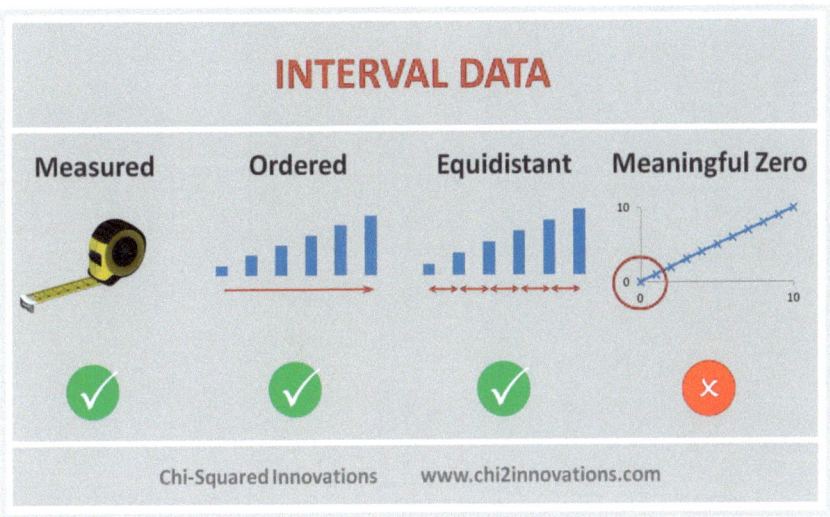

Interval data can be continuous (have an infinite number of steps) or discrete (organised into categories), and the degree of difference between items is meaningful (their intervals are equal), but not their ratio.

Mathematically you can compare the degrees of the data (equality/inequality, more/less) and add/subtract the values, such as '6pm is 3 hours later than 3pm'. However, you cannot multiply or divide the numbers, so you can't say '6pm is twice as late as 3pm'.

Ratio Data has the following properties:

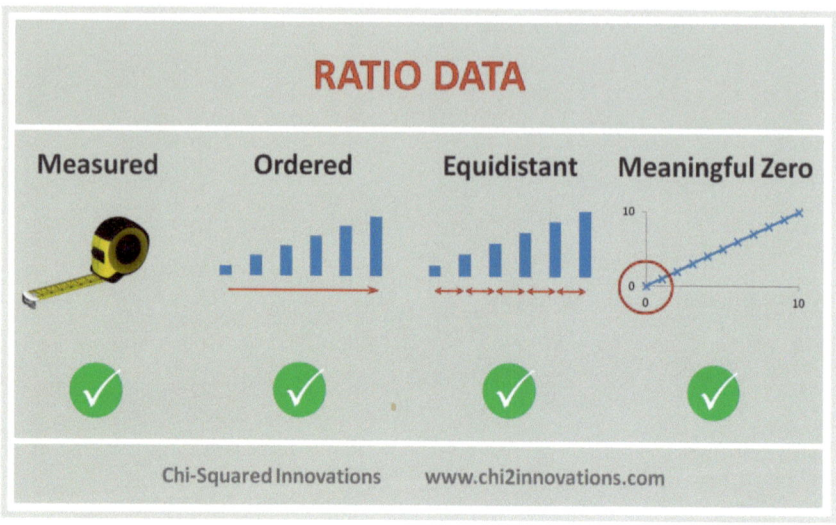

As with interval data, ratio data can be continuous or discrete, and differs from interval data in that there is a non-arbitrary zero-point to the data.

Ratio data are the best to deal with because all possibilities are on the table. You can find the central point of the data by using any of the mode, median or mean and use all of the most powerful statistical methods to analyse the data.

Knowing your data types is really powerful, because they help you to understand your data and how to analyse them. Ever heard of Nominal and Ordinal Logistic Regressions? If you have Nominal or Ordinal data, then these may be some of the statistics you might consider for your study. I hope you get the point I'm trying to make – understanding your data types from even before you've started collecting data will help you decide exactly which types of statistical analyses you'll perform on your data. It is this level of planning that will avoid the need for your statistician to scrap your 3-year long study because it wasn't thought through properly and your data are not fit for purpose.

**

Learn More

Now I don't want to come off as being some kind of a pest. I don't want you to think I'm *any* kind of a pest, but I bet you haven't visited the **Resources page** yet have you? Hands up if you haven't visited yet…

When you get to the end of this book, you'll read my bio, and when you do, you'll discover that **my mission is to unleash your inner data ninja**.

It wouldn't be much of a mission if I couldn't get you to visit one little blog post where you can find all the resources you'll need to learn about cleaning and handling your data, would it?

Don't forget, there's all sorts of FREE stuff there for you to find, so pop on over – I'll see you when you get there:

- http://bit.ly/2Rb06Ds

Chapter 4: Data Integrity

A man who has committed a mistake and does not correct it

is committing another mistake

Confucius

***

Just because you've got a perfectly clean, classified, codified and organised dataset, it doesn't mean that the data are correct.

Real life follows rules, and your data must too !!!

I once discovered that we had the oldest man in the world currently being treated in the hospital.

At well over 300 years old he'd clearly had 'a good innings'.

In the dataset I was analysing, the difference between his date of birth (somewhere in the 18th century) and date of hospital admission (21st century) meant that he was very old indeed.

Or perhaps his DOB wasn't quite right...

The error in his DOB couldn't be detected by standard error-checking in Excel because it was a perfectly legitimate date.

Tip #19: Check That Your Data Are Sensible

Sometimes, putting together 2 or more pieces of data can reveal errors that otherwise can be difficult to find, so it is sensible to **do a few simple calculations** on each variable to check that the data conform to sensible rules, such as:

- Calculate the minimum, maximum and mean
- Keep a count for each variable and each category
- Check differences between dates

Making these checks (in a separate worksheet!) lets you find outliers, such as people who have a negative age or are several hundred years old, and gives you **a good feel for your data**.

	A	B	C	D
1		Gender	Age (y)	Height (m)
2	Count	2105	2002	2212
3	Minimum	0	-312.3	1.31
4	Mean	0	53.2	1.73
5	Maximum	0	93.6	19.53
6	Negatives	27	32	0
7	Zeros	15	0	12

Other things to check include the number of negative values and zeros for each variable. Most errors can be found with just a few descriptive statistics, so learn to use them on your data.

Something doesn't feel right about the answers?

Then dive back in and take a look.

There really is no substitute for getting your hands dirty!

***

Chapter 5: Work Smarter, Not Harder

***

Bonus Tip: Automate Your Data Cleaning

Even if you've followed all of the tips here, it will still take you days or weeks to clean your dataset – and that's if it's small.

Cleaning large datasets can take months or longer.

Wouldn't it be great if you could **clean your data automatically** in minutes rather than weeks or months?

We think so, which is why **this is exactly what we've done**.

We've created a fully automated data cleaning tool – **DataKleenr** – that is:

- ✓ **Fast**
- ✓ **Simple**
- ✓ **Accurate**

Better still, it is *intelligent*, so the more data it cleans the faster and more accurate it becomes.

And you can even **use it for FREE:**

- ✓ **Save time AND money**
- ✓ **Eliminate stress**
- ✓ **Complete your research sooner**

So **check out DataKleenr**[†], then **come and talk to us**[‡].

We'd love to hear from you !!!

***

[†] https://www.chi2innovations.com/datakleenr/

[‡] https://www.chi2innovations.com/contact/

Learn More

Now we're getting near to the end of the book. So far, at the end of every chapter I've put in a reminder for you to visit the **Resources page**. I'm not going to put a reminder at the end of this chapter. I'm not. You've got the message. No more reminders to **visit the Resources page**. Not even any subliminal ones.

I'm not going to remind you of the **FREE books and video courses**. Nor of all the other FREE stuff that you can get.

Finally, I'm not going to remind you that this is the link to the resources page:

- http://bit.ly/2Rb06Ds

About The Author

Lee Baker

His mission is to unleash your inner data ninja!

***

Lee Baker is an award-winning software creator that lives behind a keyboard in a darkened room. Illuminated only by the light from his monitor, he aspires to finding the light switch.

With decades of experience in science, statistics and artificial intelligence, he has a passion for telling stories with data, yet despite explaining it a dozen times, his mother still doesn't understand what he does for a living.

Insisting that data analysis is much simpler than we think it is, he authors friendly, easy-to-understand books that teach the fundamentals of data analysis and statistics – like this one!

As the CEO of Chi-Squared Innovations, one day he'd like to retire to do something simpler, like crocodile wrestling.

Blog: https://www.chi2innovations.com/category/blog/

Newsletter: https://www.chi2innovations.com/newsletter-signup/

***

Leave a Review

Thank you for reading **Practical Data Cleaning**.

I hope you enjoyed reading it as much as I enjoyed writing it. If you did, please take a moment to leave a review. The best reviews will be featured at the beginning of the book.

Thank you!

Lee Baker